ULTIMATE BOOK FOR ROOFING

I0479951

Practical self guiding book to roof completely

Charlotte O. Harris

Table of Contents

CHAPTER ONE

INTRODUCTION TO ROOFING

Roof is a substantial part to building. A roof is made up of three essential layers, the structural deck, the underlayment, and the roofing material.

Deck features

The deck, moreover referred to as decking or sheathing, used to be traditionally made from planks of secure lumber, then again spherical the core of the twentieth century, builders swiftly transitioned to plywood or OSB oriented strand board sheets, even though some roofing materials slate, wood shingles, tile then again use lumber planks that are gapped at unique intervals referred to as pass by sheathing to promote airflow.

The variety of roof decking you have is not continually terribly important, and then again the health of it is. Roof decking ought to be flat, rigid, and free of damage and rot to furnish a robust foundation for the roofing material.

Underlayment features

Roof underlayment covers the decking and acts as a secondary wind and moisture barrier sincerely a backup for the roofing material.

Until recently, most roofs had tar paper additionally viewed as roofing felt underlayment, alternatively pretty a range synthetic elements made of woven plastic fibers have come to be extra and extra standard. Synthetic underlayment's are commonly lengthier lasting and lightweight and are less complicated to work with than roofing felt. Underlayment is hooked up in an overlapping pattern; masses like roof shingles, to aid shed any moisture that receives by the roofing material.

Roofing materials features

The two biggest enemies of roofs are time and acts of nature, which helps provide a rationalization for why few homeowners get away roof alternative forever.

Time interprets to put on and tear and the progressively ravaging penalties of sunlight, rain, snow, wind, debris, moss, critters, basketballs, and even foot traffic. Acts of nature are heaps swifter and more devastating: massive hail, immoderate winds, and falling timber or tree limbs. Roofs can moreover be destroyed by way of fire, of course, and most roofing supplies elevate a furnace rating, then again this is increased about now no longer catching on furnace from flying embers, etc. than about surviving a dwelling fire.

Strong nature of roofing in preventing winds present

As roofs age, they lose their protective coatings such as minerals on asphalt shingles and emerge as weaker and more inclined to damage. Eventually, they emerge as a criminal accountability and have to be modified to maintain the integrity of the home and to prevent high-priced problems, like water leaks. If a roof's life is minimize short, it is commonly due to storm damage, most typically hail or immoderate winds.

CHAPTER TWO

ROOFING GUIDE

Steps In Roofing Asphalt Shingles

Make edges in alignment to board inches

Drip edges are a quintessential section of any roof. Water has an immoderate ground tension, which functionality that rain tends to hold close to your roof as it rolls down the peak. Without a drip edge, water will curl up under the aspect of your roof, essential to leaks and water damage over time. You have to install the drip component before than you set up the shingles thinking about it sits on the fascia board. You choose the drip section to be shut to the phase of the fascia board then again no longer touching.

Felt paper and ensure all angle is cover

Leave about a half-inch gap between the bottom of the drip edge's kicker and the fascia board. Attach the drip section with 1–1/4–inch roofing nails. Before you start placing in the shingles, you moreover desire to cowl the roof with felt paper. Standard 30-pound felt paper is a well-known choice that works well in most climates. The felt paper's job is to structure a buffer between the shingles and the roof sheathing. This is fundamental so that your roof and shingles can prolong and contract independently when the temperature changes.

Nail your felt paper

Without the felt paper, your shingles will bind to your roof and won't have the flexibility to adapt when the local weather changes. Felt paper moreover helps take in condensation that may also additionally shape below your shingles. Attach the felt paper with the same roofing nails you used to exercise the drip edges.

Flat lays and make exterior vents

Once again, precision is now no longer necessary; in reality make positive they felt lays flat toward the roof barring any ridges or folds. You can additionally have to exchange the flashing spherical your chimney and any exterior vents you have if it seems worn or damaged. However, most human beings get away with reusing their roof's contemporary flashing. Take care when getting rid of shingles spherical your chimney, skylights, vents, and siding to hold away from detrimental the flashing.

Trim and remove previous shingles

If it seems good, put it to the side for reuse later. Before inserting in your new shingles, you favor to dispose of your modern-day shingles, nails, and underlayment. The exceptional way to do this is with a roofing shovel, a specially-designed machine that's available to wedge under shingles and makes the job lots faster. Be cautious spherical sensitive areas like your siding and chimney, usually if you graph on reusing the flashing. You may additionally desire to forge off stubborn shingles by hand with a hammer. Make positive to scan your roof for stray felt fasteners and nails, as these favor to come off formerly than placing in the new shingles.

Clean up for easy nailing and place of shingles

Sweep your roof free of any particles that stays with a stiff-bristled broom. Replacing your shingles is the perfect opportunity to furnish your roof deck an once-over. Look for signs of water damage, unevenness, warping, or damage. Unfortunately, you'll have to pause the shingle set up to restoration any issues find, then again your future self will thank you. Once you're positive the entirety seems good, you can commence inserting in the drip edges, following the approach noted in the previous section.

Nail the drip edges and avoid overlap

How you nail the drip edges isn't as necessary as making certain you have them aligned efficaciously and spaced away from the fascia board with the resource of about one-half inch. This step can be complicated for human beings barring roofing experience, so take your time if you've with the aid of no capability hooked up felt paper before.

Measure the width and length

The reason is to have an even overlaying between your shingles and roof sheathing. If you have a big roof, comply with the tips on the underlayment packaging for how to stagger the sections if one strip isn't prolonged enough to cowl the total measurement of your roof. Don't rush. Make positive to overlap the strips by using about half of their width, and be cautious no longer to introduce any folds or wrinkles as you go.

Place the shingles in rows accordingly

This is the usually most time-consuming step for new roofers, and you ought to get it right. The first row of shingles on event regarded as the starter strip has to overhang the drip phase via about a half-inch. Make sure to measure the height of your shingles and mark a line on the underlayment the area the pinnacle side of the shingles desires to be to get the half-inch overlap.

Make the vital joints

You'll in all possibility have to trim your starter route to get the tar strip as shut to the component of the roof as possible. This step will make inserting in the leisure of the shingles a lot easier, so don't pass it. Use your chalk line to snap out a grid with six-inch horizontal spacing and five-inch vertical spacing. This will make it uncomplicated to nail down the leisure of your shingles even as making certain you get the appropriate extent of overlap and exposure. When you commence inserting in the last shingles, you'll unexpectedly recognize that you'll favor to minimize the shingles to attain the appropriate horizontal offset. Some human beings pick to decrease all at as quickly as beforehand as they start, at the same time as others choose out to minimize as they go.

Nail the shingles and trim

It truly doesn't matter, so choose out whichever approach is much less intricate for you. Nail the shingles to the roof with your roofing nails, following the instructions on the shingle packaging. In general, larger nails can be sincerely beneficial if you continue to be in a region with immoderate winds.

Cut the shingles in fitting sizes

However many nails you use, make sure the subsequent shingles overlap the nail heads thru at least one inch. Unfortunately, there's no trick to inserting in shingles spherical vents, chimneys, and skylights. Cut the shingles to size as first-rate you can and use a bit of roofing cement to tightly closed them to any odd-shaped edges. Make sure no longer to go away any spaces, generally spherical any ridge vents. Making cutouts can be tricky, so having a few higher shingles on hand in case you butcher a few is an awesome idea. Clean-up, least is to installation the ridge shingles. How to install them depends upon on the manufacturer, so comply with the recommendations furnished with your purchase.

Ridge and finished up

The ordinary approach is to limit the shingles into a piece prolonged adequate to overlap the pinnacle course on each factor of the roof via 7 inches. The built-in tar strip want to be oriented for the duration of the ridge for the ridge cap shingles, perpendicular to the route it faces on the rest of the shingles. After you're finished bask in the glory of nailing down the closing shingle.

How To Roof Clay Tile

Instructions procedures

Make a roofing plan

An unstable roof physique is now not completely difficult to tile on the other hand can moreover be very dangerous. This is why you ought to make sure the structure is clearly secure and sturdy with no indication of damage. Look out for the following warning symptoms and signs when inspecting your roof structure, Lift out crucial repairs or replacements beforehand than tiling. This may additionally incorporate fixing leaks, altering elements of the bushes structure, doing away with debris, or dealing with rot.

Have a Membrane intact

The predominant attribute of roofing underlay is to furnish a greater layer of protection in the direction of moisture and exterior weathering. On older properties, you'll usually hit upon a layer of bolstered roofing felt below the tiles or slates. In ultra-modern years however, breathable roof membranes have flip out to be a loads higher well-known option.

Lay and nail from back of membrane

First, take a look at that your roof rafters are in reality free of splinters, cracks and free nails. Any leftover will favor to be eradicated to maintain away from tearing the membrane. Starting at the backside lay the first membrane roll at some stage in the trusses. Secure one give up the use of 28mm galvanized clout nails, then gently pull the leisure of the roll during to the unique side. Make high quality the membrane is laid evenly, then again now not too tightly. This can be carried out through skill of leaving reasonable sag between each rafter.

Place the battens in aligning to tiles

After accomplishing the liked tautness, restore the one-of-a-kind issue of the membrane the utilization of larger nails. Also be high quality to restoration a few on the rafters in between too. Laying your roof battens efficaciously is each different crucial step when tiling a roof. This step when executed precise will allow you to acquire the correct, even spacing between your roof tiles for most aesthetics, performance, and longevity.

Gauge and measure tiles

This spacing between the tiles is mentioned as the gauge and depends upon on your roof's pitch as exact as the dimension of tiles you're using. Place a single tile onto one of the timbers and each different onto the tile batten below. Position fixing as this will allow you to modify spacing as required. Ensure the tiles are correctly seated and hooked onto the battens. The bottom tile desires to overhang via way of spherical 60mm over the guttering.

Make roof drainage

This will allow outstanding drainage and quit water from going over onto the soffit or fascia. Measure the distance between the pinnacles of the second batten, up to spherical 34mm from the apex. This gap will accommodate contraction and increase of the roof barring detrimental the tiles.

Make two dimensions battens

Divide this distance with the aid of capability of the gauge of the genuine two battens. Round range subsequent whole number. Divide the rounded extensive range by way of the entire distance as quickly as soon as greater to calculate the quintessential gauge for the leisure of the battens. Generally, it isn't vital to nail your battens down.

Nails and avoid overlap

This is due to the truth the weight of the tiles alongside with their fixings will hold them firmly in place. Should you want to fix your battens in place, make sure you use nails with at least 44mm of penetration into the rafters. Your tiles have to overlap one some different by way of capacity of between 70mm and 100mm, relying upon their dimension as properly as the pitch of your roof.

Place and dry fix mortar

This ought to suggest that the gauge of your roofing battens will be spherical 30-35cm, measuring from the pinnacle of one batten to the pinnacle of the one beneath it. With the underlay and roof battens efficiently in place, you can now begin turning into your roof tiles. There are two techniques generally utilized in current tiling practice: moist fixing and dry fixing. Wet fixing consists of the use of mortar and cement and is commonly used for the set up of concrete roof tiles.

After dry fix, nails in the row sections

Dry fixing makes use of typically designed nails and is commonly quicker to complete. It moreover provides you the freedom to set up in truly any local weather scenario even though we continuously endorse that you pick whether or not or now not the local weather is impenetrable for tiling or not. Generally nail every 0.32 row of tiles down, establishing from the bottom, as properly as the pinnacle row. Most roof tile producers will furnish their very personal specialized fixings for use alongside with their tiles. These will have been designed to work flawlessly with their range.

Create gutter system

If your roof components an overhanging facet identified as a verge, try to allow for a 38-50mm overhang with your tiles. This will allow any rainwater to drain down into your guttering system, and now no longer over the element of your property. It's great to quit the verge piece the utilization of the absolute great part of your tile profile, as it types the most amazing edge.

End cap for ridges and finished off

Verge tiles are designed for use alongside the edges; ridge tiles are used on the very pinnacle alongside the apex, whilst hip tiles are designed for hip areas of your roof. These tiles have to be regular in the equal way as your most important set of roof tiles, whether or not or now not they've been moist or dry fixed.

Other decorative accent tiles embody merchandise such as finials, which are typically used at the pinnacle of a roof or over porches for a more extraordinary aesthetic. End caps can be added to the ends of ridges to cowl up the provide up sections and protect the tile work from the elements even similarly than, and loosen up your roof is done.

www.ingramcontent.com/pod-product-compliance
Lightning Source LLC
Chambersburg PA
CBHW070758220526
45467CB00014B/796